2022年

全国水利发展统计公报

2022 Statistic Bulletin
on China Water Activities

中华人民共和国水利部 编

Ministry of Water Resources of the People's Republic of China

·北京·

图书在版编目（CIP）数据

2022年全国水利发展统计公报 = 2022 Statistic Bulletin on China Water Activities / 中华人民共和国水利部编. -- 北京：中国水利水电出版社，2023.11
 ISBN 978-7-5226-1918-7

Ⅰ.①2… Ⅱ.①中… Ⅲ.①水利建设－经济发展－中国－2022 Ⅳ.①F426.9

中国国家版本馆CIP数据核字(2023)第215416号

书　　名	2022年全国水利发展统计公报 2022 Statistic Bulletin on China Water Activities 2022 NIAN QUANGUO SHUILI FAZHAN TONGJI GONGBAO
作　　者	中华人民共和国水利部　编 Ministry of Water Resources of the People's Republic of China
出版发行	中国水利水电出版社 （北京市海淀区玉渊潭南路1号D座　100038） 网址：www.waterpub.com.cn E-mail：sales@mwr.gov.cn 电话：（010）68545888（营销中心）
经　　售	北京科水图书销售有限公司 电话：（010）68545874、63202643 全国各地新华书店和相关出版物销售网点
排　　版	中国水利水电出版社微机排版中心
印　　刷	北京印匠彩色印刷有限公司
规　　格	210mm×297mm　16开本　3.5印张　60千字
版　　次	2023年11月第1版　2023年11月第1次印刷
印　　数	0001—1000册
定　　价	**39.00元**

凡购买我社图书，如有缺页、倒页、脱页的，本社营销中心负责调换
版权所有·侵权必究

目 录

1 水利建设投资 …………………………………… 1
2 重点水利建设 …………………………………… 4
3 主要水利工程设施 ……………………………… 7
4 水资源节约集约利用 …………………………… 12
5 防汛抗旱 ………………………………………… 14
6 水利改革与管理 ………………………………… 16
7 水利行业状况 …………………………………… 22

Contents

I. Investment in Water Projects　26

II. Key Water Projects Construction　30

III. Key Water Structures and Facilities　32

IV. Conservation and Intensive Utilization of Water Resources　38

V. Flood Prevention and Drought Relief　40

VI. Water Management and Reform　41

VII. Current Status of the Water Sector　47

2022年是党的二十大胜利召开之年,也是我国水利发展史上具有里程碑意义的一年。一年来,各级水利部门坚决贯彻党的二十大精神和习近平关于治水的重要论述精神,认真落实党中央、国务院决策部署,完整、准确、全面贯彻新发展理念,推动新阶段水利高质量发展迈出坚实步伐。

1 水利建设投资

2022年，水利建设完成投资10893.2亿元。其中：建筑工程完成投资8491.7亿元，占78.0%；安装工程完成投资486.0亿元，占4.4%；机电设备及工器具购置完成投资286.6亿元，占2.6%；其他（包括移民征地补偿等）完成投资1628.9亿元，占15.0%。2015—2022年水利建设完成投资情况见表1。

表1　2015—2022年水利建设完成投资情况　　　　单位：亿元

按规模分	2015年	2016年	2017年	2018年	2019年	2020年	2021年	2022年
全年完成投资	5452.2	6099.6	7132.4	6602.6	6711.7	8181.7	7576.0	10893.2
建筑工程	4150.8	4422.0	5069.7	4877.2	4987.9	6014.9	5851.3	8491.7
安装工程	228.8	254.5	265.8	280.9	243.1	319.7	330.1	486.0
机电设备及工器具购置	198.7	172.8	211.7	214.4	221.1	250.0	203.6	286.6
其他（包括移民征地补偿等）	873.9	1250.3	1585.2	1230.1	1259.7	1597.1	1191.0	1628.9

在全年完成投资中，防洪工程建设完成投资3628.4亿元，占33.3%；水资源工程建设（含水资源配置工程、农村规模化供水工程

等）完成投资 4473.5 亿元，占 41.1%；水土保持及生态工程建设（含复苏河湖生态环境、地下水超采综合治理等）完成投资 1625.5 亿元，占 14.9%；其他专项工程（含机构能力建设、前期工作、水库移民、三峡后续等）完成投资 1165.8 亿元，占 10.7%，其中，三峡后续工作当年完成投资 96.89 亿元。2022 年分用途完成投资情况如图 1 所示。

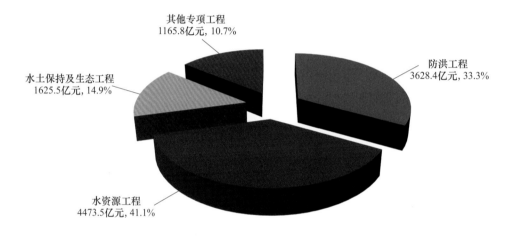

图 1　2022 年分用途完成投资情况

七大流域完成投资 9112.1 亿元，东南诸河、西北诸河以及西南诸河等其他流域完成投资 1781.1 亿元；东部、中部、西部和东北地区完成投资分别为 3587.2 亿元、3702.2 亿元、3142.6 亿元和 461.2 亿元。

在全年完成投资中，中央项目完成投资 115.5 亿元，地方项目完成投资 10777.7 亿元。大中型项目完成投资 2156.5 亿元，小型及其他项目完成投资 8736.7 亿元。各类新建工程完成投资 7947.0 亿元，扩建、改建等项目完成投资 2946.2 亿元。

当年在建的水利建设项目 40680 个，在建项目投资总规模 43210.7 亿元，较上年增加 46.5%，其中，新开工项目 25035 个，较上年增加 19.8%，新增投资规模 16580.4 亿元，是去年的 1.5 倍。2010—2022 年水利建设完成投资情况如图 2 所示。

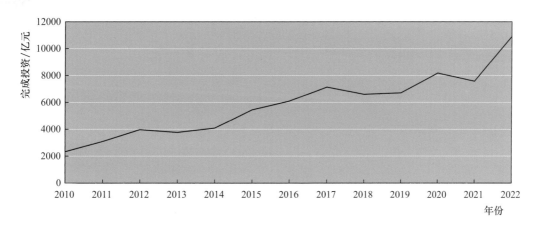

图 2　2010—2022 年水利建设完成投资情况

2 重点水利建设

流域防洪工程建设。开工建设黄河下游"十四五"防洪工程、赣江抚河下游尾闾综合整治、太湖吴淞江整治（江苏段）、淮河入海水道二期等；加快长江干流江西段崩岸应急治理、江苏重点平原洼地近期治理等重大水利工程建设；长江中下游河势控制和河道整治9项工程、黄河下游防洪2项工程全部建成并发挥效益；进一步治淮38项工程已开工33项，其中22项建成并发挥效益；洞庭湖、鄱阳湖治理5项工程已全部开工，其中4项建成并发挥效益；太湖流域水环境综合治理12项工程已全部开工，其中10项建成并发挥效益。

水资源配置工程建设。南水北调中线引江补汉、环北部湾广东水资源配置、湖北鄂北地区水资源配置二期、云南滇中引水二期、安徽引江济淮二期等工程开工建设；珠江三角洲水资源配置、重庆市渝西水资源配置、陕西引汉济渭、内蒙古引绰济辽等工程加快实施，引江济淮一期工程试通水通航，鄂北水资源配置工程顺利完工。

水库及枢纽工程建设。全年在建小型水库136座、中型水库54座、大型及枢纽工程71座。重庆藻渡、向阳，湖南大兴寨，黑龙江林海，四川青峪口，河南汉山，山东双堠、青岛官路，云南黑滩河等9座大型水库（枢纽）和福建下岩、广西板阳东等2座中型水库开工。黑龙江关门嘴子、浙江开化、河南袁湾等水库工程实现年度导截流目标；西藏湘河水利枢纽、青海蓄集峡水利枢纽、湖南毛俊等工程下闸蓄水；广东韩江高陂、江西四方井、云南车马碧等工程完工；重庆观景口、河南前坪、辽宁猴山等工程通过竣工验收。

农村水利水电工程建设。新开工江西大坳和梅江灌区、海南牛路岭灌区、广西大藤峡和龙云灌区、安徽怀洪新河灌区等8个灌区建设，继续实施大型灌区新建及现代化改造工程、中小型灌区等工程，当年新增耕地灌溉面积1227.5×10^3公顷；实施城乡供水一体化、农村规模化建设及小型工程规范化改造等工程，提升8791.1万农村人口供水保障水平，农村自来水普及率达到87%。新增农村水电站13座，新增装机15.7万千瓦。

复苏河湖生态环境工程建设。 推进福建木兰溪、吉林查干湖、安徽巢湖等一批河湖治理和生态修复。全国新增水土流失综合治理面积6.8万平方公里，其中国家水土保持重点工程新增水土流失治理面积1.31万平方公里。对622座黄土高原淤地坝进行了除险加固，整治坡耕地面积83万亩，新建淤地坝和拦沙坝790座。

3 主要水利工程设施

水库和枢纽。全国已建成各类水库95296座，水库总库容9887亿立方米。其中：大型水库814座，总库容7979亿立方米；中型水库4192座，总库容1199亿立方米。

堤防和水闸。全国已建成5级及以上江河堤防33.2万公里❶，累计达标堤防25.2万公里，堤防达标率为76.1%；其中，1级、2级达标堤防长度为3.8万公里，达标率为85.8%。全国已建成江河堤防保护人口6.4亿人，保护耕地4.2×10^3万公顷。全国已建成流量为5立方米每秒及以上的水闸96348座，其中大型水闸957座。按水闸类型分，分洪闸7621座，排（退）水闸17158座，挡潮闸4611座，引水闸13066座，节制闸53892座。2010—2022年已建成5级及以上江河堤防长度如图3所示。

---

❶ 2011年以前各年堤防长度含部分地区5级以下江河堤防长度。

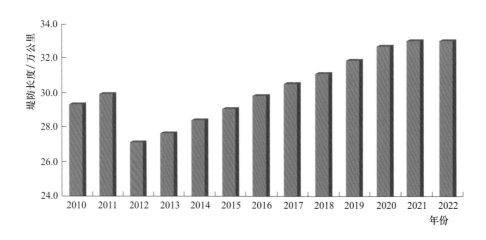

图 3　2010—2022 年已建成 5 级及以上江河堤防长度

机电井和泵站。全国已累计建成日取水大于等于 20 立方米的供水机电井或内径大于等于 200 毫米的灌溉机电井共 522 万眼。全国已建成各类装机流量 1 立方米每秒或装机功率 50 千瓦以上的泵站 94030 处，其中：大型泵站 482 处，中型泵站 4745 处，小型泵站 88803 处。

灌区工程。全国已建成设计灌溉面积 2000 亩及以上的灌区共 21619 处，耕地灌溉面积 39727×10^3 公顷。截至 2022 年年底，全国灌溉面积 79036×10^3 公顷，耕地灌溉面积 70359×10^3 公顷，占全国耕地面积的 55.1%。2010—2022 年全国耕地灌溉面积如图 4 所示。

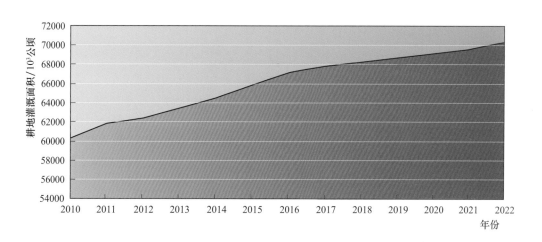

图 4　2010—2022 年全国耕地灌溉面积

农村水电站。截至 2022 年年底,全国现有农村水电站 41544 座,装机容量 8063 万千瓦,占全国水电装机容量的 19.4%。全国农村水电年发电量 2360 亿千瓦时,占全口径水电发电量的 19.6%。2010—2022 年全国农村水电站装机容量如图 5 所示。

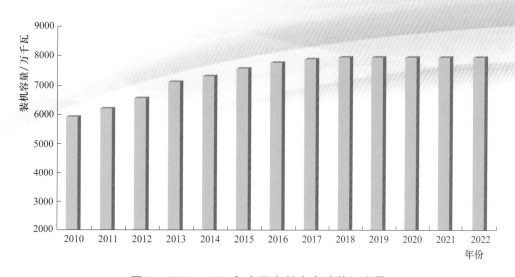

图 5　2010—2022 年全国农村水电站装机容量

水土保持工程。全国水土流失综合治理面积达156万平方公里❶，累计封禁治理保有面积达30.6万平方公里。2022年持续开展全国全覆盖的水土流失动态监测工作，全面掌握县级以上行政区、重点区域、大江大河流域的水土流失动态变化。2010—2022年水土流失治理面积如图6所示。

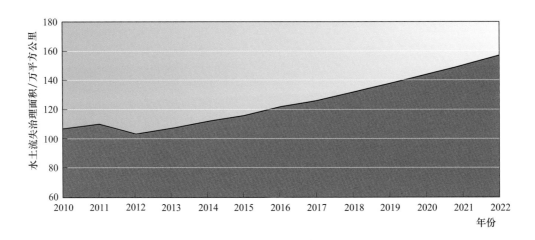

图6　2010—2022年水土流失治理面积

水文站网。全国已建成各类水文测站121731处，包括国家基本水文站3312处、专用水文站4751处、水位站18761处、雨量站53413处、蒸发站9处、地下水站26586处、水质站9737处、墒情站5102处、实验站60处。其中，向县级以上水行政主管部门报送水文信息的各类水文测站77837处，可发布预报站2630处，可发布预警站2233

---

❶ 2012年数据与第一次全国水利普查数据进行了衔接。

处；配备在线测流系统的水文测站3042处，配备视频监控系统的水文测站6470处。基本建成中央、流域、省级和地市级共331个水质监测（分）中心和水质站（断面）组成的水质监测体系。

水利网信。截至2022年年底，全国省级以上水利部门配置累计各类服务器9528台（套），形成存储能力45.96拍字节，存储各类信息资源总量达5.92拍字节；县级以上水利部门累计配置各类卫星设备3176台（套），利用北斗卫星短文传输报汛站达8169个，应急通信车140辆，集群通信终端2629个，宽、窄带单通信系统369套，无人机2157架。全国省级以上水利部门各类信息采集点达49.97万处，其中：水文、水资源、水土保持等采集点约27.47万个，大中型水库安全监测采集点约22.5万个。

4 水资源节约集约利用

水资源状况。2022年，全国水资源总量27088.1亿立方米，比多年平均值偏少1.9%。全国年平均降水量631.5毫米，比多年平均偏少2.0%，较上年减少8.7%。全国753座大型水库和3896座中型水库年末蓄水总量4180.7亿立方米，比年初减少406.2亿立方米。

水资源开发。2022年，新增规模以上水利工程❶供水能力56.0亿立方米。截至2022年年底，全国水利工程供水能力达8998.4亿立方米，其中：跨县级区域供水工程632.4亿立方米，水库工程2456.7亿立方米，河湖引水工程2114.8亿立方米，河湖泵站工程1850.8亿立方米，机电井工程1382.9亿立方米，塘坝窖池工程372.6亿立方米，非常规水资源利用工程188.2亿立方米。

水资源利用。2022年，全年总供水量5998.2亿立方米，其中：地

---

❶ 规模以上水利工程包括：总库容大于等于10万立方米水库、装机流量大于等于1立方米/秒或装机容量大于等于50千瓦的河湖取水泵站、过闸流量大于等于1立方米/秒的河湖引水闸、井口井壁管内径大于等于200毫米的灌溉机电井和日供水量大于等于20立方米的机电井。

表水供水量4994.2亿立方米，地下水供水量828.2亿立方米，其他水源供水量175.8亿立方米。全国总用水量5998.2亿立方米，其中：生活用水量905.7亿立方米，工业用水量968.4亿立方米，农业用水量3781.3亿立方米，人工生态环境补水量342.8亿立方米。与上年比较，用水量增加78.0亿立方米，其中：生活用水量减少3.7亿立方米，工业用水量减少81.2亿立方米，农业用水量增加137.0亿立方米，人工生态环境补水量增加25.9亿立方米。

水资源节约。全国人均综合用水量为425.0立方米，农田灌溉水有效利用系数0.572，万元国内生产总值（当年价）用水量49.6立方米，万元工业增加值（当年价）用水量24.1立方米。按可比价计算，万元国内生产总值用水量和万元工业增加值用水量分别比2021年下降1.6%和10.8%。全国非常规水源利用量175.8亿立方米，其中：再生水利用量150.9亿立方米，集蓄雨水利用量10.5亿立方米，淡化海水利用量4.0亿立方米，微咸水利用量3.2亿立方米，矿坑（井）水利用量7.2亿立方米。

5 防汛抗旱

2022年，洪涝灾害直接经济损失1288.99亿元（水利工程设施直接损失319.12亿元），占当年GDP的0.11%。全国农作物受灾面积$3413.73×10^3$公顷，绝收面积$492.65×10^3$公顷，受灾人口3385.26万人，因灾死亡143人，失踪28人，倒塌房屋3.13万间。江西、福建、广东、广西、辽宁、湖南等省（自治区）受灾较重。全国因山洪（山洪泥石流）灾害死亡失踪119人，占全国因洪涝灾情死亡失踪人数的69.6%。全国农作物因旱受灾面积$6090×10^3$公顷，绝收面积$612×10^3$公顷，直接经济总损失512.85亿元❶。全国因旱累计有542万城乡人口、332万头大牲畜发生临时性饮水困难。全年完成抗旱浇地面积$14268×10^3$公顷，抗旱挽回粮食损失157亿公斤，解决了521万城乡居民和251万头大牲畜因旱临时饮水困难。

---

❶ 2022年洪涝灾害直接经济损失、全国农作物受灾面积、绝收面积、受灾人口、因灾死亡和失踪人口、倒塌房屋数量，农作物因旱受灾面积、绝收面积、直接经济损失等数据来源于应急管理部。

全年中央下拨水利救灾资金83.96亿元,其中:用于防汛16.96亿元,用于抗旱67.00亿元。中央水利救灾资金在安全度汛隐患排查整治、防洪工程设施水毁修复、应急抗旱保供水等方面发挥了重要作用,为保障防洪安全、饮水安全、粮食安全提供了有力支撑。

 水利改革与管理

河湖长制。全国31个省（自治区、直辖市）党委和政府主要负责同志担任省级总河长，带领30万名省、市、县、乡级河长湖长全年巡查河湖663万人次，90万名村级河湖长（巡河员、护河员）守护河湖"最前哨"。全面建立南水北调工程河湖长制体系，设立省、市、县、乡四级河长湖长1150人，设立村级河长湖长2638人。深入推进河湖"清四乱"常态化规范化，全国共清理整治"四乱"问题2.92万个。开展全国河道非法采砂专项整治行动，累计查处非法采砂行为5839起，查扣非法采砂船舶488艘、挖掘机具1334台，移交公安机关案件179件，对全国31个省（自治区、直辖市）155个县级行政区域的河湖长、河长制办公室、水行政主管部门以及1000个河段（湖片）开展河湖长制落实情况监督检查。

水资源管理。2022年批复14条跨省江河水量分配方案；组织完成171个跨省重要河湖、指导各省完成415个重点河湖生态流量保障目标制定，16个省（自治区、直辖市）印发了地下水取水总量控制、水位控制"双控"指标；开展取用水管理专项整治行动，累计登记已建和

在建取水口 591.10 万个，涉及取水项目 84.90 万个。对黄河流域 13 个地表水超载地市、62 个地下水超采县暂停新增取水许可。2022 年全国新发取水许可电子证照 15.38 万套。推进南水北调受水区地下水压采，累计压采地下水 68.02 亿立方米。中国水权交易所 2022 年完成水权交易 3507 单，交易水量 2.5 亿立方米。

节约用水管理。2022 年，以县域为单元全面开展节水型社会达标建设，复核发布第 5 批 349 个节水型社会达标县（市、区、旗）。修订发布钢铁、纺织染整等 9 项高耗水行业用水定额国家标准。开展规划和建设项目节水评价 8130 个。推动实施合同节水管理项目 151 项。新建成水利行业节水型单位 1833 家，节水型高校 360 所，节水型灌区 182 处，遴选发布用水产品水效领跑者 30 个，重点用水企业、园区水效领跑者 74 家。推动黄河流域和京津冀地区 4 万余家年用水量 1 万立方米及以上工业和服务业单位实现计划用水管理全覆盖。

水资源调度。实施黄河、黑河、汉江等42条跨省江河流域水资源统一调度，黄河实现连续23年不断流，黑河东居延海实现连续18年不干涸，通过实施珠江枯水期水量调度保障了澳门、珠海等地供水安全。永定河首次实现春秋两次全线流动，西辽河总办窝堡枢纽实现自2002年以来首次过水，塔里木河尾闾台特玛湖水面面积和湿地生态环境有效恢复，乌梁素海水生态系统持续向好，白洋淀生态水位保证率达到100%。京杭大运河实现百年来首次全线水流贯通，与永定河实现百年交汇。华北地区河湖生态补水范围扩大至7个水系48条河（湖）流，累计补水70.22亿立方米，贯通河长3264公里、比2021年增加4.2倍，漳卫河水系、大清河白洋淀水系分别实现自20世纪60年代、80年代以来通过补水首次贯通入海，子牙河水系连续两年实现贯通入海。

运行管理。"十四五"以来，累计完成水库大坝安全鉴定34695座，完成小型水库除险加固7471座，完成小型水库雨水情监测设施建设26583座，完成小型水库大坝安全监测设施建设17667座，实行小型水库专业化管护48226座，合理妥善实施水库降等2437座、报废554座。2022年，全国786处水利工程通过省级或流域标准化管理评价，完成水利工程标准化水利部评价21处，完成大中型水闸安全鉴定1134座，完成水库、堤防、水闸确权划界3970座、22335公里、6123座。截至2022年年底，累计批准国家级水利风景区921个，其中：水库型384个，自然河湖型206个，城市河湖型211个，湿地型47个，灌区型34个，水土保持型39个。

水价改革。修订出台《水利工程供水价格管理办法》《水利工程供水定价成本监审办法》,推动建立健全有利于促进水资源节约、水利工程良性运行、与水利投融资体制改革相适应的水价形成机制。截至2022年年底,累计实施农业水价综合改革面积7.5亿亩,其中2022年新增农业水价综合改革面积1.7亿亩。

水土保持管理。全国共审批生产建设项目水土保持方案9.52万个,涉及水土流失防治责任范围3.06万平方公里;4.58万个生产建设项目完成水土保持设施自主验收报备。开展覆盖全国范围生产建设项目水土保持遥感监管,通过卫星遥感解译,组织现场核查,共认定并查处"未批先建""未批先弃"等违法违规项目1.46万个。开展国家水土保持示范创建,共评定102个示范;推动江西赣州、陕西延安、福建长汀、山西右玉、黑龙江拜泉等5个市县开展全国水土保持高质量发展先行区建设。

农村水利水电管理。截至2022年年底,25个省份累计创建绿色小水电站示范964座,长江经济带小水电清理整改完成,全国3.4万座小水电站落实生态流量。积极推进农村水电站安全生产标准化建设,全国累计建成安全生产标准化电站4700座,其中一级电站124座、二级电站1706座、三级电站2870座。

水利监督。全年共派出检查组1500个、6100人次,检查项目

12000个，就发现的各类突出问题，组织对166家责任单位实施了"约谈"及以上责任追究，其中对流域防洪工程和国家水网重大工程开展稽察5批次，派出58个稽察组、539人次；对政府质量监督履职3批次，派出30个巡查组、168人次。组织各地区各单位对28.7万处水利设施和涉及度汛工作的在建水利工程安全生产开展全覆盖自查自纠。全年有138家单位通过一级标准化达标创建评审。

依法行政。推动出台《中华人民共和国黄河保护法》。水利部（包括部机关和各流域管理机构）共受理行政审批事项78692件，办结76816件。全国立案查处水事违法案件2.06万件，当年结案1.90万件，结案率92.4%；水利部共办结行政复议案件19件，办理行政应诉18件。

水利科技。2022 年，立项实施国家重点研发计划"长江黄河等重点流域水资源与水环境综合治理""重大自然灾害防控与公共安全"等涉水重点专项项目共 16 项，国家自然科学基金长江、黄河水科学研究联合基金项目 22 项，水利技术示范项目 41 项。截至 2022 年年底，水利部共有国家和部级重点实验室 32 个（含筹建中部级重点实验室 19 个），国家和部级工程技术研究中心 15 个，国家和部级野外科学观测研究站 7 个。印发 2022 年度成熟适用水利科技推广清单，发布成果 106 项。发布水利技术标准 6 项，由水利部推荐的 4 项标准荣获 2022 年度标准科技创新奖，2 名个人荣获领军人才奖。

国际合作。2022 年，持续与芬兰、丹麦、日本等国开展政策对话与技术交流，积极参与第九届世界水论坛等国际水事活动，举办水利多双边交流活动 41 场。实施国际合作项目 71 个，8 个项目纳入国家"一带一路"建设重点项目，组织立项 15 个援外培训项目。习近平主席访问哈萨克斯坦期间，与哈方签署《中华人民共和国政府和哈萨克斯坦共和国政府关于共同管理和运行苏木拜河联合引水工程的协定》。完成对周边国家 69 个水文站国际报汛工作。

7 水利行业状况

水利单位。截至2022年年底，从事水利活动的各类县级及以上独立核算的法人单位26381个，从业人员87.3万人。其中：机关单位3045个，从业人员13.0万人；事业单位18500个，从业人员47.0万人；企业3375个，从业人员26.4万人；社团及其他组织1461个，从业人员0.9万人。

职工与工资。全国水利系统从业人员75.5万人，其中，全国水利系统在岗职工72.3万人。在岗职工中，部直属单位在岗职工5.6万人，地方水利系统在岗职工66.7万人。全国水利系统在岗职工工资总额为863.9亿元，年平均工资为12.0万元。2012—2022年职工与工资情况见表2。

表2　2012—2022年职工与工资情况

项　目	2012年	2013年	2014年	2015年	2016年	2017年	2018年	2019年	2020年	2021年	2022年
在岗职工人数/万人	103.4	100.5	97.1	94.7	92.5	90.4	87.9	82.7	77.8	74.8	72.3
其中：部直属单位/万人	7.4	7.0	6.7	6.6	6.4	6.4	6.6	6.6	6.7	6.0	5.6
地方水利系统/万人	96.0	93.5	90.4	88.1	86.1	84.0	81.3	76.1	71.1	68.8	66.7
在岗职工工资/亿元	389.1	415.3	451.4	529.4	640.5	739.1	802.7	787.6	790.9	818.7	863.9
年平均工资/(元/人)	37692	41453	46569	55870	69377	83534	91307	95236	102000	110000	120000

表3 全国水利发展主要指标（2017—2022年）

指标名称	单位	2017年	2018年	2019年	2020年	2021年	2022年
1. 灌溉面积	10^3 公顷	73946	74542	75034	75687	78315	79036
2. 耕地灌溉面积	10^3 公顷	67816	68272	68679	69161	69609	70359
其中：本年新增	10^3 公顷	1070	828	780	870	1114	1228
3. 万亩以上灌区	处	7839	7881	7884	7713	7326	
其中：30万亩以上	处	458	461	460	454	450	
4. 万亩以上灌区耕地灌溉面积	10^3 公顷	33262	33324	33501	33638	35499	
其中：30万亩以上	10^3 公顷	17840	17799	17994	17822	17868	
5. 农村自来水普及率	%	80	81	82	83	84	87
6. 除涝面积	10^3 公顷	23824	24262	24530	24586	24480	24129
7. 水土流失治理面积	万平方公里	125.8	131.5	137.3	143.1	149.6	156.0
其中：本年新增	万平方公里	5.9	6.4	6.7	6.4	6.8	6.8
8. 水库	座	98795	98822	98112	98566	97036	95296
其中：大型水库	座	732	736	744	774	805	814
中型水库	座	3934	3954	3978	4098	4174	4192
9. 水库总库容	亿立方米	9035	8953	8983	9306	9853	9887
其中：大型水库	亿立方米	7210	7117	7150	7410	7944	7979
中型水库	亿立方米	1117	1126	1127	1179	1197	1199
10. 全年水利工程总供水量	亿立方米	6043	6016	6021	5813	5920	5998
11. 堤防长度	万公里	30.6	31.2	32.0	32.8	33.1	33.2
保护耕地	10^3 公顷	40946	41409	41903	42168	42192	41972
堤防保护人口	万人	60557	62837	67204	64591	65193	64284
12. 水闸总计	座	103878	104403	103575	103474	100321	96348
其中：大型水闸	座	892	897	892	914	923	957
13. 年末全国水电装机容量	万千瓦	34168	35226	35804	37028	39094	41350
全年发电量	亿千瓦时	11967	12329	13021	13553	13399	12020

续表

指 标 名 称	单位	2017 年	2018 年	2019 年	2020 年	2021 年	2022 年
14. 农村水电装机容量	万千瓦	7927	8044	8144	8134	8290	8063
全年发电量	亿千瓦时	2477	2346	2533	2424	2241	2360
15. 当年完成水利建设投资	亿元	7132.4	6602.6	6711.7	8181.7	7576.0	10893.2
按投资来源分：							
（1）中央政府投资	亿元	1757.1	1752.7	1751.1	1786.9	1708.6	2217.9
（2）地方政府投资	亿元	3578.2	3259.6	3487.9	4847.8	4236.8	6004.1
（3）国内贷款	亿元	925.8	752.5	636.3	614.0	698.9	1450.7
（4）利用外资	亿元	8.0	4.9	5.7	10.7	8.1	5.9
（5）企业和私人投资	亿元	600.8	565.1	588.0	690.4	718.2	1065.5
（6）债券	亿元	26.5	41.6	10.0	87.2	104.3	75.1
（7）其他投资	亿元	235.9	226.3	232.8	144.9	101.1	74.0
按投资用途分：							
（1）防洪工程	亿元	2438.8	2175.4	2289.8	2801.8	2497.0	3628.4
（2）水资源工程	亿元	2704.9	2550.0	2448.3	3076.7	2866.4	4473.5
（3）水土保持及生态建设	亿元	682.6	741.4	913.4	1220.9	1123.6	1625.5
（4）水电工程	亿元	145.8	121.0	106.7	92.4	78.8	107.3
（5）行业能力建设	亿元	31.5	47.0	63.4	85.2	79.3	124.4
（6）前期工作	亿元	181.2	132.0	132.7	157.3	136.5	730.0
（7）其他	亿元	947.5	835.8	757.4	747.3	793.8	2217.9

说明：1. 本公报不包括香港特别行政区、澳门特别行政区及台湾省的数据。
2. 农村水电的统计口径为单站装机容量 5 万千瓦及以下的水电站。

2022 STATISTIC BULLETIN ON CHINA WATER ACTIVITIES

Ministry of Water Resources, P. R. China

2022 is the year when the 20th National Congress of the Communist Party of China was successfully held, and also a milestone year in the history of water resources development in China. Over the past year, the State Council, water resources departments at all levels have put the important concepts of the 20th National Congress of the Communist Party of China, and Xi Jinping's water governance philosophy, and carried forward implementation of decisions and arrangements of the Central Committee of the Communist Party of China and the State Council; pursued the new development concept in a complete, accurate, and comprehensive manner; and taken solid steps in promoting high-quality water resources development in the new stage.

I. Investment in Water Projects

In 2022, the total investment in water projects amounted to 1089.32 billion Yuan, among which, 849.17 billion Yuan was being allocated for construction projects, accounting for 78.0%; 48.60 billion Yuan for installation, accounting for 4.4%; 28.66 billion Yuan for expenditure on purchases of machinery, electric equipment and instruments, accounting for 2.6%; and 162.89 billion Yuan for other purposes, including compensation for resettlement and land acquisition, accounting for 15.0%. The Table 1 is about completed investment of water project 2015 – 2022.

Table 1 Completed investment of water project 2015－2022

unit：billion Yuan

	2015	2016	2017	2018	2019	2020	2021	2022
Total completed investment	545.22	609.96	713.24	660.26	671.17	818.17	757.60	1089.32
Construction project	415.08	442.20	506.97	487.72	498.79	601.49	585.13	849.17
Installation project	22.88	25.45	26.58	28.09	24.31	31.97	33.01	48.60
Purchase of machinery, equipment and instruments	19.87	17.28	21.17	21.44	22.11	25.00	20.36	28.66
Others (including compensation of resettlement and land acquisition)	87.39	125.03	158.52	123.01	125.97	159.71	119.10	162.89

In the total completed investment, 362.84 billion Yuan was allocated to the construction of flood control projects, accounting for 33.3%; 447.35 billion Yuan was for the construction of water resources projects, included water resources allocation projects and rural large-scale water supply projects etc, accounting for 41.1%; 162.55 billion Yuan was for soil and water conservation and ecological restoration (including revitalizing the ecological environment of rivers and lakes and comprehensive treatment of groundwater overexploitation etc), accounting for 14.9%; and 116.58 billion Yuan for specific projects (including capacity building、reservoir resettlement, the follow-up work of the Three Gorges Project etc), accounting for 10.7%. The Figure 1 is about Completed investment of various projects in 2022.

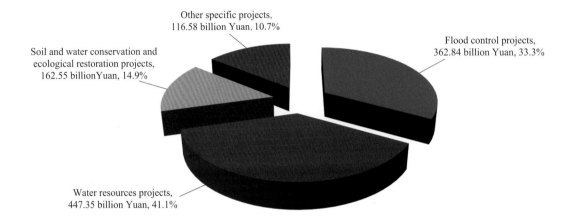

Figure 1　Completed investment of various projects in 2022

The competed investment for seven major river basins reached 911.21 billion Yuan, of which 178.11 billion Yuan was invested in river basins in the southeast, northwest and southwest of China, while investments in river basins in east, central, west and northeast China were 358.72 billion Yuan, 370.22 billion Yuan, 314.26 billion Yuan and 46.12 billion Yuan, respectively.

Of the total competed investment, the Central Government contributed 11.55 billion Yuan, and local governments contributed 1,077.77 billion Yuan. The investments completed by large and medium-sized projects reached 215.65 billion Yuan; and investment completed by small and other projects reached 873.67 billion Yuan. The investments for new projects were 794.70 billion Yuan, and investments completed by rehabilitation and expansion projects were 294.62 billion Yuan.

A total of 40,680 water projects were under construction in 2022, with a total investment of 4,321.07 billion Yuan and an increase of 46.5% over the previous year. There were 25,035 new projects launched in 2022, with an increase of 19.8% and an increase of investment of 1,658.04 billion Yuan that was 1.5 times over the previous year. The Figure 2 is about completed investments of water projects 2010–2022.

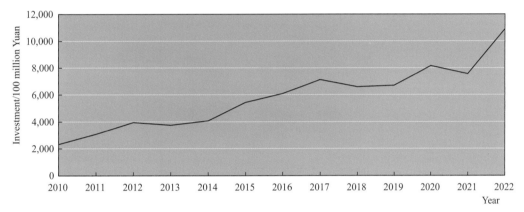

Figure 2　Completed investments of water projects 2010–2022

II. Key Water Projects Construction

Basin flood control projects. Following projects started construction, including flood control projects at the downstream of the Yellow River in the 14th Five Year Plan, integrated governance for the lower reaches of the Fuhe River in Ganjiang River, improvements of Wusong River in the Taihu Lake (Jiangsu section), and Phase-II of waterway of the Huaihe River entering into the sea. Key projects such as emergency treatment of bank collapse in the Jiangxi section of the Yangtze River mainstream and short-term plan for improvements of lowlands of key plains in Jiangsu. The construction of 9 projects for river regime regulation and river course training and restoration in the middle and lower reaches of the Yangtze River were completed for benefit generation. 2 flood control works in the lower reaches of the Yellow River completed construction for benefit generation. Out of the 38 Huaihe River improvement projects, 33 started construction and 22 put into operation and generated benefits. 5 restoration projects for Dongting and Poyang lakes started construction and 4 of which were completed for benefit generation. A total of 12 projects for Comprehensive Improvement of Water Environment of Taihu Lake began construction, and 10 of which completed for benefit generation.

Water allocation projects. In 2022, the projects started construction include water diversion from the Yangtze to the Hanjiang of the Middle Route of South-to-North Water Diversion Project, water allocation around the Beibu Gulf in Guangdong Province, second phase of water allocation in northern part of Hubei Province, second phase of water diversion to the middle part of Yunnan Province, and second phase of water diversion from the Yangtze to the Huaihe River in Anhui Province. Progressed sped up for the projects of water allocation in the Pearl River Delta, water transfer to the western part of Chongqing, water diversion from the Hanjiang to the Weihe and water diversion from Zhuoer River to Xiliaohe. Trial opening of water and navigation for the first phase of water diversion from the Yangtze to the

Huaihe River in Anhui Province. The first phase of water allocation in northern part of Hubei Province was completed.

Reservoir and water control projects. In 2022, construction commenced for 136 small reservoirs, 54 medium-sized reservoirs and 71 large reservoirs. The large reservoirs include Zaodu Reservoir and Xiangyang Reservoir in Chongqing, Daxingzhai Reservoir in Hunan, Linhai Reservoir in Heilongjiang, Qingyukou Reservoir in Sichuan, Hanshan in Henan, Shuanghou Reservoir and Guanlu Reservoir in Shandong, and Heitanhe Reservoir in Yunnan. 2 medium-size reservoirs were started construction, namely Xiayan Reservoir in Fujian and Banyangdong Reservoir in Guangxi. Damming was completed for Guanmenzuizi Reservoir in Heilongjiang, Kaihua in Zhejiang and Yuanwan in Henan. Impoundments were completed for water storage of Xianghe Multipurpose Dam Project in Xizang, Xujixia Multipurpose Dam Project in Qinghai, Maojun Reservoir in Hunan. Construction of Gaopi Multipurpose Dam Project in Hanjiang of Guangdong, Sifangjing Multipurpose Dam Project in Jiangxi and Chemabi Reservoir in Yunnan were completed. Guanjingkou Multipurpose Dam Project in Chongqing, Qianping Reservoir in Henan and Houshan Reservoir in Liaoning passed check and acceptance for benefit generation.

Rural water supply, irrigation and hydropower development. Construction of rural water resources and hydropower projects. 8 new irrigation districts, including Jiangxi Da'ao and Meijiang Irrigation District, Hainan Niululing Irrigation District, Guangxi Datengxia and Longyun Irrigation District, and Anhui Huaihong Xinhe Irrigation District started construction. The construction and modernization of large-scale irrigation districts, as well as small and medium-sized irrigation districts had been carried out continuously. In 2022, the newly-increased irrigated area of cultivated land reached 1,227,500 hm^2; and Projects such as the integration of urban and rural water supply, large-scale construction in rural areas, and rehabilitation of small projects for standardization had been implemented the guarantee rate of water supply increased to 87,911 million rural population. The percentage of rural population access to tap water rose to 87%. The 13 new hydropower stations increased installed capacity 157,000 kW.

Restoration of river and lake ecological environment. A number of projects of river and lake governance and ecological restoration had been proceeded, including Mulan Creek in Fujian, Chagan Lake in Jilin, and Chaohu Lake in Anhui. In 2022, the newly-increased areas for soil and water conservation and comprehensive control of soil erosion reached 68,000 km^2, of which the areas under the National Major Project for Soil Conservation amounted to 13,100 km^2. Up to 622 silt-retention dams on Loess Plateau at high risk were strengthened and rehabilitated. The slope farmland of 830,000 mu were harnessed. There were 790 newly-built silt retention dams and check dams.

III. Key Water Structures and Facilities

Reservoirs and dams. The reservoirs built in China were 95,296, with a total storage capacity of 988.7 billion m^3. Of which 814 are large reservoirs, with a total capacity of 797.9 billion m^3 and 4,192 reservoirs are medium-sized with a total capacity of 119.9 billion m^3.

Embankments and water gates. The completed river dykes and embankments at Grade-V or above reached 332,000 km[1]. The accumulated length of dykes and embankments that met the standard reached 252,000 km, accounting for 76.1% of the total. Among which, up-to-the-standard dykes and embankments of Grade-I and Grade-II reached 38,000 km, taking up 85.8% of the total. The completed dykes and embankments nationwide could protect 640 million people and 42,000 hm^2 of farmland. The completed water gates with a flow of 5 m^3/s were 96,348, of which 957 were large water gates. By type, there were 7,621 flood diversion sluices, 17,158 drainage/return water sluices, 4,611 tidal barrages, 13,066 water diversion intakes and 53,892 controlling gates. The Figure 3 is about Length of completed river dykes and embankments at Grade-V or above 2010–2022.

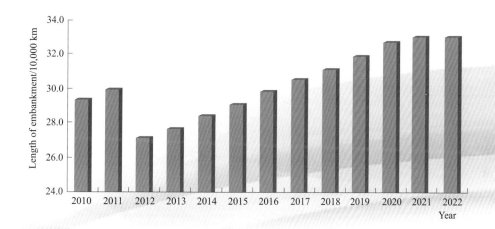

Figure 3 Length of completed river dykes and embankments at Grade-V or above 2010–2022

---

[1] The length of river dykes in a year before 2011 included the length of river dykes below Grade-V in some regions.

Tube wells and pumping stations. There were 5.22 million tube wells with a daily water abstraction capacity equal to or larger than 20 m³ or an inner diameter equal to or larger than 200 mm, completed for water supply in the whole country. A total of 94,030 pumping stations with a flow of 1 m³/s or an installed voltage above 50 kW were put into operation, including 482 large, 4,745 medium-size and 88,803 small pumping stations.

Irrigation schemes. Irrigation districts with a designed area of 2,000 mu or above were 21,619 in total, covering 39.727 million hm² of irrigated farmland. By the end of 2022, the total irrigated area amounted to 79.036 million hm². The irrigated area of cultivated land reached 70.359 million hm² that accounted for 55.1% of the total in China. The Figure 4 is about Irrigated area of cultivated land 2010-2022.

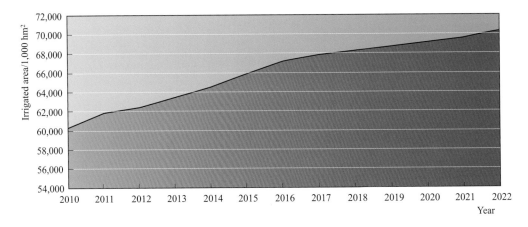

Figure 4   Irrigated area of cultivated land 2010-2022

Rural hydropower and electrification. By the end of 2022, the number of hydropower stations built in rural areas was 41,544, with an installed capacity of 80.63 million kW, accounting for 19.4% of the national total. The annual power generation by these hydropower stations reached 236.00 billion kW·h, accounting for 19.6% of the national total. The Figure 5 is about Installed capacity of rural hydropower 2010–2022.

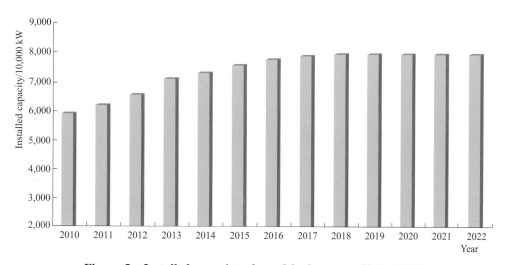

Figure 5 Installed capacity of rural hydropower 2010–2022

Soil and water conservation. The area with soil conservation measures reached 1.56 million km² ❶; and the forbidden area for ecological restoration accumulated to 306,000 km². In 2022, dynamic monitoring had been continued for soil and water loss in all administrative areas above county level, key areas, and major river basins in the country, in order to collect comprehensive data on dynamic changes. The Figure 6 is about The area recovered from soil erosion 2010–2022.

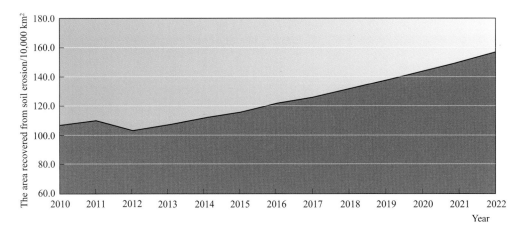

Figure 6 The area recovered from soil erosion 2010–2022

Hydrological station networks. There were 121,731 hydrological stations of all kinds constructed in the whole country, including 3,312 national basic hydrologic stations, 4,751 special hydrologic stations, 18,761 gauging stations, 53,413 precipitation stations, 9 evaporation stations, 26,586 groundwater monitoring stations, 9,737 water quality stations, 5,102 soil moisture monitoring stations and 60 experimental stations. Among them, 77,837 stations of various kinds can provide hydrological information to water administration authorities at and above county level; 2,630 stations can release forecast and 2,233 can release early

---

❶ The data in 2012 had been integrated with the data of the First National Water Resources Survey.

warnings; 3,042 equipped with online flow measurement and 6,470 equipped with video monitoring system. A water quality monitoring system consisted of 331 monitoring centers, sub-centers and water quality stations (sections) at central, basin, provincial and local levels was set up.

Water networks and information systems. By the end of 2022, water resources departments and authorities at and above provincial level were equipped with 9,528 servers of varied kinds, forming a total storage capacity of 45.96 PB, and keeping 5.92 PB of data and information. Water resources departments and authorities at and above county level had equipped with 3,176 sets of various kinds of satellite equipment, 8,169 flood forecasting stations for short message transmission from the Beidou Satellites, 140 vehicles for emergency communication, 2,629 cluster communication terminals, 369 narrowband and broadband communication systems, and 2,157 unmanned aerial vehicles (UAV). A total of 499,700 information gathering points were available for water resources departments and authorities at and above county level, including 274,700 points for collecting data of hydrology, water resources and soil and water conservation and 225,000 points for safety monitoring at large and medium-sized reservoirs.

IV. Conservation and Intensive Utilization of Water Resources

Water resources conditions. In 2022, the total water resources in the country amounted to 2,708.81 billion m³, 1.9% less than the normal years. The mean annual precipitation was 631.5 mm, 2.0% less than the normal years and 8.7% less than the previous year. The total storage of 753 large and 3,896 medium-sized reservoirs were 418.07 billion m³, reducing 40.62 billion m³ comparing to that at the beginning of the year.

Water resources development. In 2022, the newly-increased capacity of water supply by water structures above designated size❶ was 5.6 billion m³. By the end of 2022, the total capacity of water supply reached 899.84 billion m³, among which 63.24 billion m³ was from water supply systems at the county level, 245.67 billion m³ from reservoirs, 211.48 billion m³ from water diversion of rivers and lakes, 185.08 billion m³ from pumping stations of rivers and lakes, 138.29 billion m³ from electro-mechanical wells, 37.26 billion m³ from ponds, weirs and cellars, and 18.82 billion m³ from unconventional water sources.

Water resources utilization. In 2022, the total quantity of water supply amounted to 599.82 billion m³, including 499.42 billion m³ from surface water, 82.82 billion m³ from groundwater and 17.58 billion m³ from other sources. The total water consumption amounted to 599.82 billion m³, including 90.57 billion m³ of domestic

---

❶ Water projects above designated size include reservoirs with a total capacity of 100,000 m³ or higher, pump stations with an installed flow at or above 1 m³/s or an installed capacity at or above 50 kW, water diversion gates with a flow at or above 1 m³/s, electric irrigation wells 200 mm or larger in inner diameter or with a water supply capacity at or above 20 m³/d.

water use was, 96.84 billion m³ of industrial water use was, 378.13 billion m³ of agricultural water use and 34.28 billion m³ for environment and ecological use. Comparing to the previous year, the total water consumption increased 7.80 billion m³, among which agricultural water use increased 13.70 billion m³ and artificial recharge for environmental and ecological use increased 2.59 billion m³; however, domestic water use decreased 370 million m³ and industrial water use decreased 8.12 billion m³.

Water saving and conservation. The mean annual water consumption was 425 m³ in China. The coefficient of effective use of irrigated water was 0.572. Water use per 10,000 Yuan of GDP (at comparable price of the same year) decreased to 49.6 m³ and that per 10,000 Yuan of industrial value added (at comparable price of the same year) reduced to 24.1 m³. Based on the estimation at comparable prices, water uses per 10,000 Yuan of GDP and per 10,000 Yuan of industrial value added reduced by 1.6% and 10.8% respectively over the previous year. The consumption of unconventional water sources amounted to 17.58 billion m³, among which 15.09 billion m³ from reclaimed water, 1.05 billion m³ from collected or stored rainwater, 0.40 billion m³ from desalinated water, 0.32 billion m³ from blackish water and 0.72 billion m³ from treated mine water.

V. Flood Prevention and Drought Relief

In 2022, the direct economic loss caused by flood and waterlogging disasters was 128.899 billion Yuan (including 31.912 billion Yuan of direct losses of water facilities), accounting for 0.11% of the GDP in the same year. A total of 3,413,730 hm^2 of cultivated land were affected by floods, including 492,650 hm^2 non-harvest farmland, affected population of 33.8526 million, 143 dead and 28 missing and 31,300 collapsed houses. The provinces suffered heavily from severe flooding included provinces/autonomous region of Jiangxi, Fujian, Guangdong, Guangxi, Liaoning and Hunan provinces (autonomous regions). The death toll and missing of mountain floods totaled 119 that accounted to 69.6% of the total national number as a result of flood and waterlogging. The affected areas of farmland by drought were 6,090,000 hm^2 and areas with no harvest were 612,000 hm^2, with a total of 51.285 billion Yuan of direct economic loss[1]. A total of 5.42 million urban and rural residents and 3.32 million man-feed big animals and livestock suffered from temporary drinking water shortage. Up to 14,268,000 hm^2 of lands were irrigated against droughts that retrieved a grain loss of 15.7 billion kg. Drinking water was provided to 5.21 million rural and urban population and 2.51 million big animals and livestock in order to alleviate temporary water shortage.

In, 2022, the Central Government allocated 8.396 billion Yuan for the mitigation of water-related disasters, including 1.696 billion Yuan for flood defense and 6.700 billion Yuan for drought relief. The funds for disaster relief from the Central Government have played a key role in ensuring safety during the flood season, hazard investigation and mitigation, repairing of damaged structures and facilities

---

[1] The data of direct economic losses caused by floods, the affected area of crops, the area of no harvest, the affected population, the number of dead and missing due to disasters, and the number of collapsed houses in 2022 all come from the Ministry of Emergency Management.

and emergency water supply, and provide strong support for safeguarding flood control and drinking water security and grain security.

VI. Water Management and Reform

River (lake) chief system. In 2022, the governors in 31 provinces, autonomous regions and municipalities, served as the general chiefs, took the lead in patrol and investigation of rivers and lakes for 6.63 million men-times, together with 300,000 river/lake chiefs at provincial, city, county and township levels. More than 900,000 river/lake chiefs (including those for river patrol and protection) at village level acted as the "outposts" to guide rivers and lakes. The river/lake chief system had been established for the South-to-North Water Diversion Project, to nominate 1,150 chiefs at provincial, city, county and township levels. There were 2,638 chiefs at the village level were nominated in line with local conditions. Normalization and standardization had been applied to avoid "misappropriation, illegal sand excavation, disposing of wastes and building structures without permission", with a total of 29,200 illegal activities corrected. A special action was taken against illegal sand-mining, to investigate and give punishment on 5,839 illegal sand-mining cases and captured 488 vessels and 1,334 excavating machines, and turned over 179 illegal cases to the public security. Supervision and inspection had been conducted of for the implementation of river and lake chief system in 155 county-level administrative regions of 31 provinces (autonomous regions, municipalities directly under the central government) across the country including river and lake chief system offices, water administrative departments, and 1000 river sections (lake area) in 2022.

Water resources management. Formulation of quotas was accelerated to strengthen water resources management, with approval of 14 cross-province water allocation plans. The targets were specified for securing ecological flow of 171 cross-province major rivers and lakes. Guidelines were adopted by provinces to define the targets of securing ecological flow of 415 major rivers and lakes. Indicators of "double control" on total groundwater use and water level control for 16 provinces/autonomous regions/municipalities were issued and made public. A special action of "looking back" nationwide involved register of 5.911 million constructed and under-constructed water intakes that relate to 849,000 water abstraction projects. The issue of new water abstraction licenses was suspended for 13 prefectures and cities with surface water over-abstraction and 62 counties with groundwater over-abstraction. In 2022, the newly issued electronic licenses reached 153,800 in the whole country. More effects had been made on reduction of groundwater withdrawal in water receiving areas of South-to-North Water Diversion Project, with an accumulated reduction of 6.802 billion m^3 of groundwater withdrawal. China Water Exchange had completed a total 3,507 entitlement trading, with an amount of 250 million m^3 of water.

Water conservation management. In 2022, the construction of up-standard water-saving society at the county level was continued with approval of 349 counties or districts in 5 batches. There were 9 national standards for water use quota of high water consumption industries issued. Water-saving evaluation was conducted for 8,130 plans and construction projects. Contract-based water conservation and management was extended with 151 projects implemented. In the construction of water-saving carrier pilots, 1,833 institutions and organizations in the water sector were awarded the title of water-saving organization; 360 universities and colleges were awarded as water-saving universities; 182 irrigation districts were awarded as water-saving irrigation districts. There were 30 products were selected and titled "water efficiency leader". 74 enterprisers and industrial parks were also awarded the title of "major water consuming enterprises" or "park water efficiency leader". More emphasis had placed on planed water use that covered 40,000 enterprises of industrial and service sectors in the Yellow River Basin and Beiing-Tianjin-Hebei Region, that have an annual water consumption of 10,000 m^3 and above, so as to realize full coverage management of planned water use.

Integrated regulation of water resources in river basins. Integrated d regulation of water resources had been implemented in river basins of the Yellow River, Heihe River, Hanjiang River and other 42 trans-provincial rivers. The Yellow River had been flowing continuously for 23 consecutive years, and the Dongjuyanhai Lake at the lower reaches of the Heihe River had not been dried up for 18 consecutive years. Water supply security of Macao, Zhuhai and other cities had been guaranteed through the implementation of water regulation in the dry season of the Pearl River. Ecological regulation in key areas of a river basin has been continued, as the Yongding River has achieved full line flowing in both spring and autumn for the first time. Water was released by Zongbanwopu Multi-purpose Dam Project in the Xiliao River Basin for the first time since 2002. Water surface and wetland ecological environment of Taitema Lake at the tail end of the Tarim River have been restored effectively. Water ecosystem of Wuliangsuhai continues to improve. Water diversion from the Yellow River to Hebei has maximized benefits to ensure a 100% compliance rate for the ecological water level of the Baiyangdian Wetlands. Water replenishment was operated for the entire line of the Beijing-Hangzhou Grand Canal, which helped to realize full-line water conveyance for the first time in a century. Ecological environment recovery of rivers and lakes in North China has been undertaken which expanded the scope of ecological water replenishment to 7 water systems and 48 rivers/lakes. In 2022, a total of 7.022 billion m^3 of water has been used for replenishment, equivalent to 3,264 km of river length, an increase of 4.2 times compared to the amount in 2021. The water systems of the Zhangwei and Baiyangdian of the Daqing River had realized their first connections to the sea since the 1960s and 1980s respectively due to water replenishment. Ziya River system had been connected to the sea for two consecutive years.

Operation and management. From the beginning of the 14th Five-Year Plan period until the end of 2022, safety appraisal had been implemented for 34,695 reservoirs being expired in 2020. A total of 7,471 small reservoirs had been reinforced, with 26,583 rainwater monitoring facilities installed; 17,667 dam safety monitoring facilities installed; 48,226 professionally managed and protected according to local conditions; and 2,437 reservoirs downgraded and 554 removed. In 2022, 786 projects across the country passed evaluation of standardized management at provincial or basin levels. MWR made evaluations on standardization of 21 water resources projects. Safety evaluation was applied to 1,134 large and medium-sized water gates. Authorization of water entitlement and delimitation of administrative boundaries had been completed for 3,970 reservoirs, 22,335 kilometers of embankments, and 6,123 water gates. As of the end of 2022, a total of 921 national water scenic spots of various kinds had been approved, including 384 reservoir type, 206 natural rivers and lakes, 211 urban river and lake type, 47 wetlands, 34 irrigation districts, and 39 soil and water conservation areas.

Water pricing reform. In collaboration with the National Development and Reform Commission (NDRC), MWR revised and issued *the Management Measures for Price of Water Supply of Water Resources Projects* and *the Supervision and Examination Measures for Pricing and Costs of Water Supply of Water Resources Projects*, in order to create and perfect water pricing mechanism that contribute to water conservation, healthy operation of water resources projects and reform of investment and financing system. As of the end of 2022, reform of water pricing system had been applied to 750 million mu of agricultural land accumulatively, of which 170 million mu were newly added in 2022.

Soil and water conservation. In 2022, MWR approved 95,200 soil and water conservation plans of construction projects, covering an area of 30,600 km^2 within the scope of responsible for control of soil and water losses. Soil conservation facilities of 45,800 construction projects passed self-check and acceptance. Remote sensing was adopted in the whole country to monitor soil and water conservation caused by human activities in production and construction. The on-site reinvestigations were organized according to the interpretation of data from remote sensing satellites, with 14,600 projects punished because of "construction without approval" and "damping wastes without permission". Demonstration areas were built for soil and water conservation at national level, with 102 pilots approved. High-quality development has been promoted in the pilot cities or counties, namely Ganzhou of Jiangxi Province, Yan'an of Shaanxi Province, Changting of Fujian Province, Youyu of Shanxi Province and Baiquan of Heilongjiang Province.

Rural hydropower management. By the end of 2022, 964 small hydropower stations in 25 provinces met the criteria of green small hydropower station. Removal or rehabilitation of small hydropower stations in the Yangtze River Economic Belt was completed, and ecological flow should be provided by 34,000 small hydropower stations. Standards for safe production and operation had been applied to hydropower stations in rural areas, with 4,700 hydropower stations complied with relevant standards, including 124 of level A, 1,706 of level B and 2,870 of level C hydropower stations in the whole country.

Supervision. In 2022, inspection teams of 1,500, with 6,100 person-times, were dispatched for 12,000 projects. MWR organized "regulatory talks" or more serious disciplinary actions to 166 enterprises who had failed to comply with the relevant stipulations. More attention had been paid on inspection of flood control projects in river basins and main schemes of the national water network, with 5 batches of inspection and a total of 58 inspection teams and 539 personnel. The focus had been placed on performance of government agencies for quality supervision, with dispatching of 30 inspection teams and 168 personnel to conduct inspections in 3 batches on the performance. Self inspection and correction for safety production had been conducted for 287,000 water structures and under-constructed projects related to work of flood control at the local level. In 2022, 138 organizations passed the review of first level standardization.

Law-based administrative and enforcement. In 2022, *Yellow River Protection Law of the People's Republic of China* was promulgated. In 2022, MWR handled 78,692 applications for water-related administrative approvals or permits with 76,816 completed. MWR investigated 20,600 water-related cases and resolved 19,000 cases, with a rate of 92.4%. MWR handled and concluded 19 administrative reconsideration cases and 18 administrative proceedings.

Water science and technology. In 2022, the central government approved and allocated to 16 special-subject water-related projects, such as Comprehensive Regulation of Water Resources and Water Environment in Major River Basins of the Yangtze River and the Yellow River and Prevention and Control of Major Natural Disasters and Public Security; for 22 projects with joint funds for scientific studies on the Yangtze River and the Yellow River; for 41 water science and technologies demonstration projects. By the end of 2022, MWR had 32 national level or ministerial level labs (including 19 labs in preparation stage); 15 national and ministerial level engineering technology research centers; and 7 national and ministerial level field scientific observation and research stations. The name-list of applicable water technologies for scientific extension and 106 achievements in 2022 were released. There were 6 water-related technical standards made public, 151 standards under drafting and 803 currently effective water-related technical standards. 4 standard recommended by the Ministry of Water Resources won the 2022 standard Science and Technology Innovation Award, and 2 individuals won the Leading Talent Award.

International cooperation. In 2022, MWR had continued to facilitate policy dialogue and technical exchanges with its counterparts from Finland, Denmark, Japan and other countries, and organized 41 multi-bilateral exchange activities. Delegations were sent to participated the 9th World Water Forum. There were 71 international cooperation projects under implementation, 8 of which were included in the list of national key projects of "the Belt and Road" initiative, and 15 foreign-aid training projects. During visit of President Xi Jinping to Kazakhstan, an inter-governmental agreement was signed with Kazakhstan on *joint management and operation of water diversion projects on the Sumbe River.* Flood monitoring and early warning to the neighboring countries had been completed through 69 hydrological stations in China.

VII. Current Status of the Water Sector

Water-related institutions. By the end of 2022, there were 26,381 legal entities, with 873,000 employees and separate accounts, engaged in water-related activities within the administrative jurisdiction at county level or above. Among them, the number of governmental organizations was 3,045 with 130,000 employees; public organizations of 18,500 with 470,000 employees; enterprises of 3,375 with 264,000 employees; societies and other institutions of 1,461 with 9,000 employees.

Employees and salaries. Employees of the water sector totaled 755,000. Of which, in-service staff members amounted to 723,000, including 56,000 working in agencies directly under the Ministry of Water Resources and 667,000 working in local agencies. The total salary of in-service staff members nationwide was 86.39 billion Yuan, and the annual average salary per person was 120,000 Yuan. The Table 2 is about employees and salaries 2012–2022.

Table 2 Employees and Salaries 2012–2022

Composition of Funds	2012	2013	2014	2015	2016	2017	2018	2019	2020	2021	2022
Number of in service staff /10^4 persons	103.4	100.5	97.1	94.7	92.5	90.4	87.9	82.7	77.8	74.8	72.3
Of them: staff of MWR and agencies under MWR /10^4 persons	7.4	7.0	6.7	6.6	6.4	6.4	6.6	6.6	6.7	6.0	5.6
Local agencies /10^4 persons	96.0	93.5	90.4	88.1	86.1	84.0	81.3	76.1	71.1	68.8	66.7
Salary of in-service staff /10^8 Yuan	389.1	415.3	451.4	529.4	640.5	739.1	802.7	787.6	790.9	818.7	863.9
Average salary /(Yuan/person)	37,692	41,453	46,569	55,870	69,377	83,534	91,307	95,236	102,000	110,000	120,000

Table 3　Main Indicators of National Water Resources Development 2017－2022

Indicators	Unit	2017	2018	2019	2020	2021	2022
1. Irrigated area	10^3 hm^2	73,946	74,542	75,034	75,687	78,315	79,036
2. Farmland irrigated area	10^3 hm^2	67,816	68,272	68,679	69,161	69,609	70,359
Newly-increased in 2022	10^3 hm^2	1,070	828	780	870	1,114	1,228
3. Irrigation districts over 10,000 mu	unit	7,839	7,881	7,884	7,713	7,326	
Irrigation districts over 300,000 mu	unit	458	461	460	454	450	
4. Farmland irrigated areas in irrigation districts over 10,000 mu	10^3 hm^2	33,262	33,324	33,501	33,638	35,499	
Farmland irrigated areas in irrigation districts over 300,000 mu	10^3 hm^2	17,840	17,799	17,994	17,822	17,868	
5. Rural population accessible to safe drinking water	%	80	81	82	83	84	87
6. Flooded or waterlogging area under control	10^3 hm^2	23,824	24,262	24,530	24,586	24,480	24,129
7. Controlled or improved eroded area	10^4 km^2	125.8	131.5	137.3	143.1	149.6	156.0
Newly-increased in 2012	10^4 km^2	5.9	6.4	6.7	6.4	6.8	6.8
8. Reservoirs	unit	98,795	98,822	98,112	97,072	97,036	95,296
Large-sized	unit	732	736	744	756	805	814
Medium-sized	unit	3,934	3,954	3,978	4,043	4,174	4,192
9. Total storage capacity	10^8 m^3	9,035	8,953	8,983	9,086	9,853	9,887
Large-sized	10^8 m^3	7,210	7,117	7,150	7,231	7,944	7,979
Medium-sized	10^8 m^3	1,117	1,126	1,127	1,146	1,197	1,199
10. Total water supply capacity of water projects in a year	10^8 m^3	6,043	6,016	6,021	5,813	5,920	5,998
11. Length of dikes and embankments	10^4 km	30.6	31.2	32.0	32.8	33.1	33.2

Continued

Indicators	Unit	2017	2018	2019	2020	2021	2022
Cultivated land under protection	10^3 km²	40,946	41,351	41,903	42,168	42,192	41,972
Population under protection	10^3 people	60,557	62,785	67,204	64,591	65,193	64,284
12. Total water gates	unit	103,878	104,403	103,575	103,474	100,321	96,348
Large-sized	unit	892	897	892	914	923	957
13. Total installed capacity by the end of the year	10^4 kW	34,168	35,226	35,804	37,028	39,094	41,350
Yearly power generation	10^8 kW·h	11,967	12,329	13,021	13,553	13,399	12,020
14. Installed capacity of rural hydropower by the end of the year	10^4 kW	7,927	8,044	8,144	8,134	8,290	8,063
Yearly power generation	10^8 kW·h	2,477	2,346	2,533	2,424	2,241	2,360
15. Completed investment of water projects	10^8 Yuan	7,132.4	6,602.6	6,711.7	8,181.7	7,576.0	10,893.2
Divided by different sources							
(1) Central government investment	10^8 Yuan	1,757.1	1,752.7	1,751.1	1,786.9	1708.6	2,217.9
(2) Local government investment	10^8 Yuan	3,578.2	3,259.6	3,487.9	4,847.8	4236.8	6,004.1
(3) Domestic loan	10^8 Yuan	925.8	752.5	636.3	614.0	698.9	1,450.7
(4) Foreign funds	10^8 Yuan	8.0	4.9	5.7	10.7	8.1	5.9
(5) Enterprises and private investment	10^8 Yuan	600.8	565.1	588.0	690.4	718.2	1,065.5
(6) Bonds	10^8 Yuan	26.5	41.6	10.0	87.2	104.3	75.1
(7) Other sources	10^8 Yuan	235.9	226.3	232.8	144.9	101.1	74.0

Continued

Indicators	Unit	2017	2018	2019	2020	2021	2022
Divided by different purposes							
(1) Flood control	10^8 Yuan	2,438.8	2,175.4	2,289.8	2,801.8	2,794.0	3,628.4
(2) Water resources	10^8 Yuan	2,704.9	2,550.0	2,448.3	3,076.7	2,866.4	4,473.5
(3) Soil and water conservation and ecological recovery	10^8 Yuan	682.6	741.4	913.4	1,220.9	1,123.6	1,625.5
(4) Hydropower	10^8 Yuan	145.8	121.0	106.7	92.4	78.8	107.3
(5) Capacity building	10^8 Yuan	31.5	47.0	63.4	85.2	79.9	124.4
(6) Early-stage work	10^8 Yuan	181.2	132.0	132.7	157.3	136.5	730.0
(7) Others	10^8 Yuan	947.5	835.8	757.4	747.3	793.8	2,217.9

Notes:

1. The data in this bulletin do not include those of Hong Kong, Macao and Taiwan.

2. Statistics of rural hydropower stations with an installed capacity of 50,000 kW or lower than 50,000 kW.

《2022年全国水利发展统计公报》编辑委员会

主　　　任：陈　敏
副　主　任：张祥伟
委　　　员：（按姓氏笔画排序）
匡尚富　邢援越　巩劲标　任骁军　刘宝军　孙　卫
李兴学　李春明　李原园　李　烽　杨卫忠　吴　强
张新玉　陈茂山　郑红星　赵　卫　姜　成　袁其田
夏海霞　钱　峰　倪　莉　郭孟卓　曹纪文　曹淑敏
谢义彬　靳宏强

《2022年全国水利发展统计公报》主要编辑人员

主　　　编：张祥伟
副　主　编：谢义彬　吴　强　汪习文
执 行 编 辑：张光锦　李　淼　张　岚
主要参编人员：（按姓氏笔画排序）
万玉倩　马　超　王小娜　王鹏悦　曲　鹏　刘　品
孙宇飞　杜崇玲　李天良　李云成　李笑一　杨　波
吴泽斌　吴海兵　吴梦莹　沈东亮　张晓兰　张慧萌
陈文艳　周哲宇　房　蒙　殷海波　郭　珅　郭　悦
黄藏青　蒋雨彤　韩绪博　谢雨轩　廖丽莎　潘利业
英 文 翻 译：谷丽雅

◎ 主编单位
水利部规划计划司

◎ 协编单位
水利部发展研究中心

◎ 参编单位
水利部办公厅
水利部政策法规司
水利部财务司
水利部人事司
水利部水资源管理司
全国节约用水办公室
水利部水利工程建设司
水利部运行管理司
水利部河湖管理司
水利部水土保持司
水利部农村水利水电司
水利部水库移民司
水利部监督司
水利部水旱灾害防御司

水利部水文司
水利部三峡工程管理司
水利部南水北调工程管理司
水利部调水管理司
水利部国际合作与科技司
水利部综合事业局
水利部信息中心
水利部水利水电规划设计总院
中国水利水电科学研究院